IMAGES
of America

QUONSET POINT
NAVAL AIR STATION

Dedication

Dedicated to My Beloved Family and the unsung Heroes of Naval Aviation: the Enlisted Men
of the United States Navy, Marine Corps, and Coast Guard.

Sean Paul Milligan, December 1995

(Drawing by Kate Huntington)

IMAGES
of America

QUONSET POINT
NAVAL AIR STATION

Sean Paul Milligan

ARCADIA
PUBLISHING

Published by Arcadia Publishing
Charleston, South Carolina

Printed in the United States of America

For all general information contact Arcadia Publishing at:
Telephone 843-853-2070
Fax 843-853-0044
E-mail sales@arcadiapublishing.com
For customer service and orders:
Toll-Free 1-888-313-2665

Visit us on the Internet at www.arcadiapublishing.com

Contents

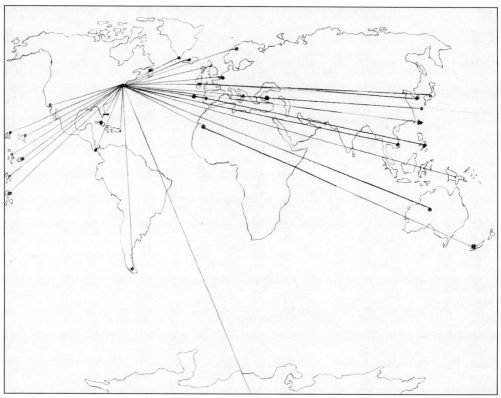

Quonset and Comfair Quonset's air squadrons, carrier air groups, fleet air wings, and aircraft carriers, plus the products of the great industrial naval air station's A&R/O&R/NARF (Naval Air Rework Facility) plant served oceans, seas, and continents worldwide. For thirty-four years, locations included the Arctic, the Antarctic, Ireland, Great Britain, Norway, Italy, Greece, Turkey, the Azores, Portugal, North Africa, France, Israel, Lebanon, Iceland, Newfoundland, Labrador, Cuba, San Diego (CA), Hawaii, Midway Island, Wake Island, the Philippines, the Carolines, the Marianas, the Marshalls, Tarawa, Tinian, Saipan, Guam, Okinawa, New Zealand, Australia, Japan, China, Argentina, Thailand, and Vietnam. (Map created by Mr. Jesse Gray Lee)

Introduction

In 1939 there were only four U.S. Naval Air Stations that could support traditional seaplane operations and provide a landing field and other facilities for aircraft carrier-type planes. Stations that could, in addition to this, berth and support four aircraft carriers, operate a modern, multi-runway airfield that could handle the heaviest four-motored planes, and maintain a technically-superior industrial plant and research and development complex, did not exist.

Two years later, there was one—Naval Air Station Quonset Point, Rhode Island.

Conceived after Nazi Germany invaded Poland, the air station was originally to be a simple seaplane base for the Neutrality Patrol of "FDR's Navy." However, plans for Quonset expanded exponentially as the war in Europe worsened and spread into the Western Atlantic. The great naval air station, built mostly on reclaimed sea bottom, was completed in record time and was considered an engineering miracle second only to the construction of the Panama Canal. More than a year before the Japanese attack on Pearl Harbor, Quonset's flying sailors were out over the North Atlantic, protecting convoys bound for England from Nazi U-boat torpedoes. Over half of all U-boats sunk by U.S. Naval Aviation, including the first two, were destroyed by shore-based and carrier-based squadrons trained and supported at Quonset.

Throughout World War II, Quonset's aviators ranged around the world, hitting the enemy everywhere, day and night: from the freezing Arctic Circle, through the warm Mediterranean, the treacherous Atlantic, the deadly Bay of Biscay, and also far westward, into the boiling Pacific, to Tokyo, and beyond, to victory. After the Tarawa campaign, at least half of all aircraft carrier formations in the Pacific had Quonset origins, and many were in Tokyo Bay for the surrender of Imperial Japan in August 1945.

After World War II, NAS Quonset Point maintained its status as the premier U.S. Industrial Naval Air Station. It was the spawning ground for the Navy's first real carrier-based jet squadron, Fighting Seventeen Able (VF-17A); the springboard for the detailed air exploration of the Antarctic; and was involved in many other top-secret scientific projects.

Peacetime activity at the air station's giant and highly-competent overhaul and repair department included remanufacturing more than two hundred R4D DC-3 transports, as well as even greater numbers of F4U Corsairs and Douglas AD Skyraiders that would see combat in Korea.

Quonset's flying sailors were among the first to see combat in Korea and were also among the very last to operate over that unfortunate country. The Korean War was a benchmark display of the ineptitude of the newly-formed United Nations, ineptitude that would lead America into another Asian war ten years later.

With the Korean conflict stalemated in 1953, NAS Quonset Point and its top command echelon, "Comfair Quonset," focused most of its efforts on the new Cold War threat to America:

attack by Russian missile submarines and aircraft, armed with nuclear weapons. By 1958, all jet attack and propeller air groups were long gone from Quonset to air stations south of the Mason-Dixon line. Quonset was thus exclusively in the ASW business; one year later, with 160-knot S2F Grumman Tracker aircraft pursuing fast Russian nuclear submarines in the North Atlantic, it was a losing business.

Quonset, as the premier U.S. Industrial Naval Air Station, got a very temporary new lease on life during the 1960s, when light and medium jet attack aircraft, plus propeller-driven types, got chewed up in Vietnam and had to be repaired at Quonset's Naval Aircraft Rework Facility.

For thirty-four years and through four wars—three hot and one cold—NAS Quonset Point, with Yankee work ethic, skill, drive, devotion to duty, and sacrifice, defended America and the free world, quite well, *and then some*.

Sean Paul Milligan
Rhode Island
December 1995

Credit Legend

MIT=Massachusetts Institute of Technology Radiation Laboratory
NA=National Archives
NHC=Naval Historical Center
PC=Providence College Archives
QAM=Quonset Air Museum
USMC=United States Marine Corps
USN=United States Navy
USNI=United States Naval Institute
USCG=United States Coast Guard

One

Construction and Early Operations

By December 23, 1940, additional hangar construction was well under way as Quonset's first skipper, and Rhode Island native, Commander Harold J. Brow made the first wheeled landing aboard the air station, in a borrowed JF-1 Grumman Duck. Three days earlier, after some sharp maneuvering around steam dredges, Brow made Quonset's first water landing in another Duck. (USN/QAM)

As early as 1918, the General Board for Naval Affairs recommended that a seaplane base be established on Narragansett Bay, either at Newport or Quonset, to protect the Northeast from German submarine and surface ship attack. Had World War I continued into 1919, twin-motored machines, such as this Curtiss H-16 (complete with a Lewis gun), would have been assigned to the area. The lofty perch between the twin V-12 Liberty engines is manned by a Marine, while the waist gun position is handled by a sailor, c. 1918. (USN/Raytheon)

Good drop! U.S. Naval Aviation's bona fide interest in Narragansett Bay started in 1918, when Gould Island was purchased for use in the development of aerial torpedoes. By 1920, a hangar and seaplane ramp were constructed to handle machines such as this Curtiss R-6L. On August 30, 1921, two aircraft were delivered to the small East Passage Island, and by late December they were installed in what would become the Navy's oldest continuous aviation unit, the Newport Torpedo Station's Air Detail. (NA)

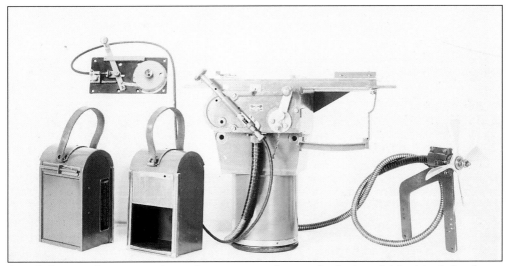

Aerial photography gear, similar to this equipment (except for the large high-altitude lens), was used by the Air Detail to record the successes and failures of aircraft torpedo drops in the East Passage Range. All manner of torpedo development was vital to the Navy at this time, in lieu of indecisive early twentieth-century results: in the Pacific between Japan and Russia in 1905, and a decade later, at the Battle of Jutland, between England and Germany. (Raytheon)

On July 21, 1927, Army Air Corps "Colonel" Charles Augustus Lindbergh Jr. stopped by Quonset's National Guard (ex-Militia) Camp Grounds with his magnificent Ryan NYP. His visit occurred shortly after his famous flight to Paris, during which he flew over a portion of Rhode Island. (QAM)

Quonset's Rhode Island Militia Camp Grounds were originally established in 1898 during the Spanish-American War. The vertical spire was a signal tower for light and semaphore signal practice, *c.* 1939. (QAM)

Within a year of the initial contract signings, construction of Quonset's Neutrality Patrol hangar (later known as Seaplane Hangar Two) was well under way. (QAM/Browning)

Army observation aircraft, similar to this Douglas 0-46A, visited the National Guard Camp Grounds during the 1930s. (Ronald W. Harrison)

During the summer of 1940, large and small state-of-the-art hydraulic dredges, as well as older steam-driven bucket dredges (veterans of the Panama Canal construction), spewed out thousands of tons of bay bottom. The dredges created over 440 acres of new land, mostly for the two main runways (16-34 and 5-23), while simultaneously fashioning deep channels for the seaplane aerodrome, the perimeter bulkhead, the 1127-foot-long Carrier Pier channel, and the turning basin for Essex Class (CV-9) ships. (USN/QAM)

Struggling through ice, the Norfolk-based PBY-2 Catalina flying boat 4-J-12 of VJ-4 awaits beaching gear off a Quonset seaplane ramp. Utility Squadron Four helped put Quonset in commission and participated in the Neutrality Patrol that President Roosevelt initiated after Germany invaded Poland in September 1939. (USN/Bowers)

Navy racing pilot Lieutenant Harold J. Brow posed with his Curtiss Navy Racer #2 on September 14, 1922. A year later, flying a similar R2C-1, he would become the fastest man in the world. Seventeen years later, as a full commander, he became Quonset's first skipper. (USNI)

This photograph, taken in the summer of 1939, shows the Navy's one and only TBD-1A Devastator seaplane. A fixture of the Newport Torpedo Station's Air Detail, she probably dropped more torpedoes for tests than all her sisters did in combat during World War II. The seaplane was last seen on the Quonset A&R scrap line, on July 23, 1943. (Scarborough/Dickey)

Filmed from a VX-2D1 "Yellow Peril," the TBD-1A drops a torpedo-discharging propellant. (NA)

Commissioning Ceremonies
☆

11:00 A. M. to 2:00 P. M.
Eastern Daylight Time

COMMISSIONING OF THE NAVAL AIR STATION
Rear Admiral Edward C. Kalbfus, U. S. Navy
Commandant Naval Operating Base
Newport, R. I.

Acceptance of Command
Commander Andrew C. McFall, U. S. Navy
Commanding Officer, Naval Air Station
Quonset Point, R. I.

Hoisting of the Colors and playing of the National Anthem

Addresses by
The Assistant Secretary of the Navy
The Honorable Ralph A. Bard
and
Distinguished Guests

SETTING OF THE WATCH
•
Retreat
•

This program will be broadcast over stations
WEAN and the Mutual Broadcasting System

Arrangements for Guests
☆

**The following arrangements have been made
for your convenience**

1. The blue ticket of admission to the Commissioning Ceremonies and to the Reception and Buffet Luncheon entitles you to pass through the Police and Sentry Lines to the grandstand and admission to the Buffet Luncheon after the commissioning.

2. The blue ticket is to aid the sentries and police in identifying you and you are requested to display it prominently to facilitate such identification. These blue tickets are non-transferable.

3. Please surrender the blue tickets at the main gate when leaving.

4. Parking facilities for cars will be arranged. Transportation about the station for inspection will be furnished.

5. Commissioning Ceremonies will require approximately forty-five minutes.

6. In the event of rain, Commissioning Ceremonies will be conducted in a Seaplane Hangar.

Respectfully,
ANDREW C. McFALL,
Commander, U.S.N., Commanding

NAS Quonset Point stands up. These Commissioning Ceremonies and Assignments for Guests were held on July 12, 1941. (USN/PC)

The first aircraft "officially" assigned to NAS Quonset Point was an obsolete SF-1 Fifi scout fighter. A known "lemon," it was quickly dispatched to Squantum Naval Reserve Air Base, where it had to make a glide landing after its engine quit. Within a few days, the old biplane was scrapped. (Grumman/Lovisolo)

This congressional delegation is en route for an inspection of the proposed Primary North East Naval Air Station site (Quonset) during a snow squall on March 17, 1939. They arrived at the Hillsgrove State Airport in one of the U.S. Navy's newest Douglas R2D-1 staff transports, a VIP DC-2. A year earlier, the Hepburn Board had concluded that the Quonset site was indeed the logical choice for a great naval air station. (USN/PC)

Vought OS2U-3 Kingfishers, like this one with a 260-pound depth bomb under its belly, flew numerous, four-hour, daylight ASW patrols from Quonset, both as land-based planes and seaplanes, on grids that extended east from Gay Head and Montauk. (USMC/Leatherneck)

On the eve of the Japanese attack on Pearl Harbor, the USS *Long Island* (AVG-1, later CVE-1), America's first escort carrier, was steaming off Narragansett Bay on a Neutrality Patrol. The converted cargo ship was searching for Nazi U-boats by using modified, obsolete, battleship/cruiser catapult float planes such as this Curtiss SOC-1A, shown here catching an early wire aboard AVG-1. (USN/Larkins)

The *Long Island*, with a combination of scout and fighter-type aircraft, became the Navy's first "Jeep Carrier" while escorting North Atlantic convoys well before America's formal entry into World War II, and led the way for all Antisubmarine Warfare (ASW) aircraft carrier activity. A Grumman Wildcat fighter is shown here launching from her oblique catapult.

Five days before the Japanese attack on Pearl Harbor, a Scouting Squadron 42 (VS-42) SB2U Vought Vindicator rolls in for a message drop on the flight deck of the USS *Ranger* while on an Atlantic Neutrality Patrol U-boat hunt. (USN/Ranger Alumni)

The first "Quonset Special" arrived at the air station on April 1, 1941, bearing more than four hundred civilian construction employees from Union Station, Providence. At the request of the Navy, the New York, New Haven and Hartford Railroad continued this service through 1943. Operating on 26 miles of track between 1942 and 1974, this was, after the New Haven, the second largest railroad in Rhode Island. (USN/QAM)

A wonderful array of jitneys also hauled workers to and from Quonset. (QAM/Browning)

Several first-rate civilian summer cottages, initially condemned by the airfield survey team, were wisely saved, skidded to new locations on the air station by dray horses and tractors, and put to use as married officers' quarters. (USN/QAM)

This 1941 photograph documents dredging and pile-driving activity at the Carrier Pier and seaplane ramps. (USN/QAM)

Quonset's Power Station supplied electricity and steam heat for the entire NAS complex. During the great Northeast Blackout of November 9, 1965, this old plant, without missing a beat, kept the air station on line while the civilian power grid quickly went dead. (USN/QAM)

Attention to colors! The first of many air squadrons to be commissioned at Quonset, Inshore Patrol Squadron Two, First Naval District (VS-2D1), stands up on October 20, 1940. They were photographed in front of the former Rhode Island Militia chow hall, which at the time was serving as a temporary administration building. (USN/QAM)

While finishing touches were applied to hangars, former civilian summer cottages skidded to the center of the airfield were put to good use as blueprint rooms, tool cribs, and machine shops in April 1941. (USN/Browning)

Before he became a vice admiral, former Chief of Navigation Captain William F. "Bull" Halsey inspected the air station and proclaimed many of its features to be the best in the Navy. He especially liked the new civilian mess hall, and described its operations as being as good or better than anything he just seen in Detroit's most modern auto plants. (USN/USNI)

The Navy's interest in the potential uses of Narragansett Bay was evident on August 8, 1924, during the first use of a mooring mast aboard a ship, when the dirigible USS *Shenandoah* (ZR-1) coupled to the USS *Patoka* (AO-9) while the converted oiler was underway off Jamestown. (USN/USNI)

On January 27, 1928, the USS *Los Angeles* (ZR-3) made a "successful" landing aboard America's third aircraft carrier, the USS *Saratoga* (CV-3), off Newport. The *Los Angeles* somehow managed to transfer passengers while taking on fuel, water, and supplies. (USN/USNI)

The USS *Ranger* Air Group was the first aircraft carrier formation to undergo training and enjoy the facilities of NAS Quonset Point. Grumman CODS (carrier on-board delivery), such as this JF-1 Duck "Ranger-6" (buno 9440), were both the first and the last aircraft types to operate from Quonset. (Grumman)

On February 27, 1941, two VS-2D1 Kingfishers of Quonset's Neutrality Patrol were bedded down in Seaplane Hangar Two under the watchful eye of Commander Brow. To the right of Grumman Duck in the background is a tour bus from the 1939 New York World's Fair, employed by the air station as a traveling security office. More than sixty Quonset civilian employees were handed over to the FBI after fingerprints and mug shots disclosed the presence of illegal aliens and felons. (USN/QAM)

Commander Brow "chats" with a mechanic taking in the warm rays of the fading winter sun through Seaplane Hangar Two's big clerestories on February 27, 1941. Brow is standing near plane No. 1 of VS-2Dl, a OS2U-2 Vought Kingfisher (buno 2193), which is adorned with a Neutrality Patrol star on its engine cowling. (USN/QAM)

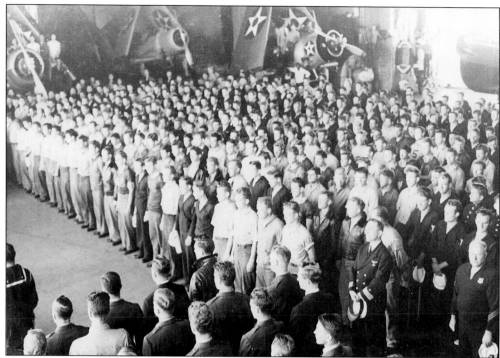

"Eternal Father strong to save . . ." All hands stood at attention for a burial at sea on the hangar deck of Quonset's first carrier, the USS *Ranger* (CV-4), on May 20, 1941. The Neutrality Patrol never came cheap to the Quonset family. (NA/Ranger Alumni)

Up from Newport, this early visitor to the great air station is a VX-2D1 Grumman J2F-1 Duck (buno 0177). (Scarborough)

Departing for Quonset from Norfolk, VP-52, PBY-5 52-P-8, with twin depth-charge racks mounted, gets a tow to the water before starting its engines. The next day, on March 26, 1941, VP-52 started escorting convoys bound for England and hunting Nazi U-boats. (Scarborough)

Arriving at Quonset, VP-52 was greeted by partially-finished hangars and ramps littered with construction debris. The seaplane control tower can be seen above Seaplane Hangar Two, near PBY-5 52-P-6. (USN/QAM)

These VP-52 PBY-5s are being launched from a temporary wooden seaplane ramp in May 1941. They relocated to the "son of Quonset" NAS Argentia, in Newfoundland, as part of "FDR's Navy." This squadron's PBY-5s helped the British track down the RKM *Bismarck*, the Nazi super battleship with 15-inch guns that ran loose in the North Atlantic after sinking the HMS *Hood*. (Scarborough)

This big, permanent, concrete seaplane ramp was photographed in May 1941 while still under construction. Seaplane Hangars One and Two, with PBY flying boats in between, had been completed by this time. (Scarborough)

Two

World War II

The ramp in front of Land Plane Hangar One is packed full with Carrier Air Group Sixteen Hellcat fighters, Avenger torpedo bombers, and Dauntless dive bombers for the "new Lex" (the USS *Lexington* CV-16). This photograph was taken from the heavily sand-bagged control tower. (NA)

This view is of the front office of the Navy's first land-based patrol bomber, the Lockheed PBO-1 Hudson. Quonset's VP-82 was the first Navy squadron to operate these machines, was the first American air unit to sink Nazi U-boats during World War II, and got to transmit home the famous, "Sighted sub, sank same." The warship-type compass is wisely located below the pilot's control yoke; "B" and "G" on the yoke spoke indicate bombs and guns; and the bumbershoot-looking handles at the top center are .303 machine-gun cocking levers. (USN/Irish)

VP-82's PBO-1s were in fact Lend-Lease Lockheed Hudson MK IIIs, with full British armament and ammo, pulled right off the assembly lines at Van Nys a few days after Pearl Harbor. These efficient land-based planes started a revolution in patrol plane aviation, which had previously relied on big flying boats, seaplanes, and amphibians. By 1968, all U.S. Navy patrol planes would be land-based and built by Lockheed. (USN/Irish)

PV-3s were retired a year later and were replaced by PV-1s. Quonset's flying sailors picked them up at Burbank, beautifully-adorned with patriotic cartoon characters like "Kid Vega" here. Unfortunately, beautiful artwork was immediately painted out upon arrival at Quonset on the orders of a visiting "flag type." Quonset's Fleet Air Wing Nine operated PVs throughout the war. Not one convoy ship was lost to U-boats during World War II while under the watchful eye of Quonset's flying sailors. (Lockheed)

Rhode Island's leading newspaper always got it right. Within hours of the *Providence Sunday Journal* hitting the streets, torpedo planes, dive and horizontal bombers, plus Zero fighters (the best in the world at that time), hit a sleeping Pearl Harbor. A "second Pearl Harbor" hit Quonset right away as more than half of the air station's PBY aircraft and crews were dispatched to Hawaii to replace those caught on the ground at Ford Island. (Providence Public Library)

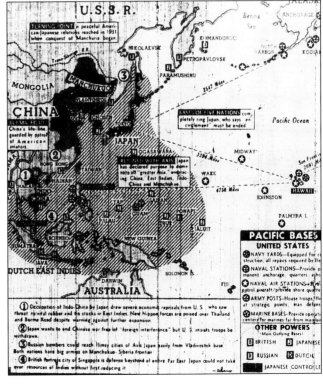

After training at Quonset, the Navy's first TBF-1 Grumman Avengers speed off westward on June 4, 1942, just in time to participate in the Battle of Midway with Torpedo Squadron Eight. Except for the shot-up remains of this one airplane (8-T-1, buno 00380) that Ensign Albert K. Earnest somehow brought back, with a dead turret gunner and wounded radioman, Torpedo Eight was wiped out. (USN/Lawson)

This 1942 photograph shows the driver's seat of a Grumman Avenger. (Grumman/Lovisolo)

One of NAS Quonset's outlying fields, Charlestown, became a full-fledged NAAS during the war and was the center for naval night-fighter training. "Charlie's" ramp is shown here in 1945, packed full with F6F Grumman Hellcats. (Webster)

Two Brewster SB2A-4 Buccaneers, similar to this Marine SB2A-2, were employed by Quonset's Project AFIRM for radar training. (USMC/Curry)

MIT's Spraycliff Observatory, located near Beavertail Point, Jamestown, was photographed in July 1945. Quonset's top secret radar installation, code named MICKEY and NAVY WHITE, was where most of the radar and CIC (Combat Information Center) equipment and tactics that helped the Navy win World War II was developed. (MIT Radiation Laboratory)

Trained in Quonset for the invasion of Southern France, this 6F-5 Hellcat revs up for cat shot and a fighter-bomber mission from the USS *Kasaan Bay* (CVE-69) on August 15, 1944. (USN/Lawson)

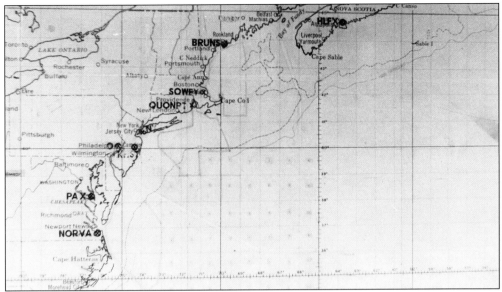

This is the NAS Quonset Point War Room Grid. Quonset's North Atlantic World War II responsibilities ranged as far east as Iceland and as far south as Bermuda. Comfair (Commander, Fleet Air) Quonset's administrative and operational authority began on January 1, 1943. (Author)

NAS Quonset Point, the "Pensacola of the North," logged in many flight mishaps and accidents, both aboard the air station and in and around Rhode Island. Unfortunately, many planes—such as this North American SNJ advanced trainer—were reduced to this state. (USN/Webster)

Curtiss P-40E Warhawks on the *Ranger* head to North Africa for General Patton's and General Eisenhower's Operation TORCH Invasion, which took place on April 15, 1942. (USN)

One of the *Ranger's* F4F Wildcats is shown here being launched during the Invasion of North Africa. (USN)

Displaying giant Neutrality Patrol stars on all six positions and equipped with a flare pod under its port wing, this Grumman F4F Wildcat was part of the *Ranger*'s Fighting Squadron Forty-One. (USN)

The crew of the "Carolyn," an Army-originated B-24D/PB4Y-1 Liberator, was photographed after "graduating" from Quonset's Fairwing Nine. (NA)

A later PB4Y-1, built for the Navy, heads to the Bay of Biscay, right where the U-boats broke out into the Atlantic. The Emerson bow turret, and all other gun stations, got plenty of use as the Bay was a real hot spot, defended by formations of cannon-armed JU-88 heavy fighters and flak ships. (USN/Fairwing Seven)

This PB4Y-1 (buno 32169) of Fairwing Seven's VB-114 suffered a gear failure on July 5, 1944, shortly after arriving in England. Fortunately, the crew was alright. Back at Quonset, another PB4Y-1 crew was not as lucky when the ship flew into Narragansett Bay while practicing night searchlight tactics. (USN/Fairwing Seven)

This one-of-a-kind PB4Y-1 Liberator, armed with a .50 caliber chin turret and retractable rocket rails mounted in the aft bomb bay, was developed by Quonset's ASDEV. Highly-respected and capable, ASDEVRONLANT's skipper, Commander Aurelius B. Vosseler, had to turn down this well-intentioned ship due to the difficulties it experienced with the high-speed/low-altitude attack modes used to destroy U-boats. (USN/Haus)

Two destroyers nest alongside the *Ranger*, while an old friend, the heavy cruiser USS *Tuscaloosa* (CA-37, with her second triple 8-inch gun turret swung outboard for routine ramming of her large naval rifles), shares the Quonset Carrier Pier on March 28, 1944. Often, while ferrying pursuit ships and fighter bombers across the Atlantic, the *Ranger*'s Carrier Air Group Four remained at the air station and took gunnery, dive bombing, and torpedo practice in Narragansett Bay. The *Ranger*'s sailormen, airdales, and ship's company loved both Providence and its "Twin City" neighbor, Pawtucket. (NA)

While the *Ranger* lies in Scapa Flow with the British Home Fleet, her skipper, Captain Gordon Rowe, shows Secretary of the Navy Frank Knox one of the Grumman torpedo planes that would be used against Nazi shipping during Operation LEADER. A few months after the attack, Captain Rowe was promoted to commodore with orders to Comfair Quonset, always considered one of the top rewards in Naval Aviation. (USN/Cressman)

October 4, 1943, was a rough day for the Nazis in Norway. Off the harbor of Bodo, the German freighter *Saar* is bombed and strafed by planes from Carrier Air Group Four, launched from the *Ranger* during Operation LEADER. The Air Group sank six enemy ships that day, severely damaged another four, and shot down two Luftwaffe reconnaissance bombers. (USN/Cressman)

October 4, 1943, was a rough day for the Nazis in the North Atlantic as well. The German U-336, similar to the U-118 shown here, was destroyed by a Navy PV-1 Ventura crew from a VB Squadron belonging to Quonset's Fairwing Nine. Land-based and carrier-based squadrons that were formed, trained, outfitted, and controlled by the Commander, Fleet Air Quonset, sank more than half of all the enemy submarines that were credited to Navy pilots during the Battle of the Atlantic. (USN/Lawson)

Crowded with lashed-down Avengers, Dauntless, and Wildcats, the *Ranger*'s snow-covered flight deck was photographed in 1943. (USN/Cressman)

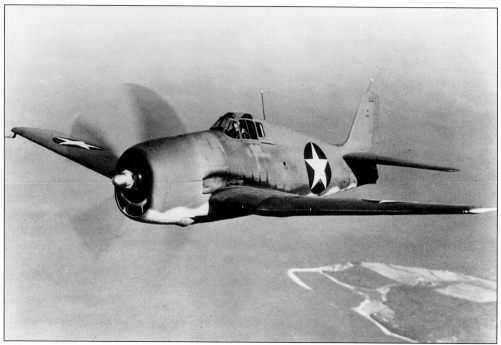

All new Grumman F6F Hellcats started their Navy life at Quonset. More Hellcats were lost in Rhode Island through training mishaps than were shot down in the Pacific in air-to-air combat with the Japanese. "Crack-a-day Quonset" became a familiar phrase. (NA/Curry)

The North American PBJ-1J was one of the rich variety of ships employed by Quonset's unique ASDEVRON (Antisubmarine Development Squadron). The one-of-a-kind PBJ-1G, armed with a 75 mm cannon, operated from Seaplane Hangar One before being turned over to the Marines. (USN)

Quonset was the birthplace of the naval night fighter. The cannon-armed Grumman F6F-5N was the fruit of Project AFIRM's labors. AFIRM became NACTU (Night Attack Training Unit) late in the war, and NAAS Charlestown became known as "Night Fighter Town USA." This photograph was taken in 1945. (USN/Webster)

In 1942, Quonset's Project AFIRM developed the world's first carrier-based night fighter, the F4U-2 Corsair. (USMC/Lawson)

This photograph, taken on April 1, 1944, shows the cockpit instrument panel of a night-fighting, APS radar-equipped, F6F-3N Grumman Hellcat, developed and perfected by Quonset's Project AFIRM. AFIRM was based in LPH-3 (Land Plane Hangar Three). (Grumman)

A Quonset sailor secures the cockpit of ASDEV TBF-1 A-14 (buno 47520), which crashed on January 16, 1944. Older, worn-out models were often used, and used up, by Quonset's top-secret development organizations. Pilots and aircrews serving in these units were always at high risk. (USN/Haus)

In this 1943 photograph, the size, girth, and ruggedness of the American-built Avengers can be seen. The man exiting the cockpit is a Royal Navy chap. (NA/Curry)

Most British Royal Navy personnel that passed through the Comfair Quonset training pipeline used Corsairs, Avengers, and Wildcats, but a "lucky" few got to check out in the Curtiss SB2C Helldiver. This is Helldiver I, JW 121, 1820 Squadron, Fleet Air Arm, Royal Navy. (Royal Navy/Curry)

An experimental MAD (Magnetic Anomaly Detection) device is shown here being dragged by an ASDEVLANT Teeb. Along with Bell Telephone and the off-shore oil-exploring Gulf Research Company, Quonset's Project SAIL and ASDEVRON developed MAD equipment to hunt U-boats. The results of this work effectively denied the Nazis access to the Mediterranean Sea by the middle of 1943. Updated, the same techniques and systems are still in use today for hunting subs. (Grumman)

This is a detail of the towed MAD gear. (Grumman)

Davisville Seabees constructed an 1,800-foot-long pontoon "aircraft carrier" for evaluation by Quonset's aviators. The crazy thing, powered by giant outboard motors, actually worked in the sheltered waters of the upper Bay, but would have been a little testy in rougher conditions. Born right next to Quonset at Davisville, the Seabees always lived up to their motto: "the difficult we do right away, the impossible takes a little longer." (USN/QAM)

These seaplane ramps were photographed in 1945. (USN)

The crews of Quonset's small, fast rescue vessels were always on alert and saved many souls. The rescue vessels were meticulously maintained and operated. (USN/QAM)

Resembling farm silos, Link trainer enclosures stand astride the Synthetic Training Building on August 24, 1944. Quonset was the first in the Navy to have these, and trained most Link operators and technicians during the war. (USN/QAM)

Construction never ended at Quonset. Brick-layers are shown here toiling at the new BOQ site on August 10, 1944. (USNI/QAM)

Rebar for new engine test cells are erected while a new carrier air group, believed to be CVG-153, with Hellcats, Avengers, Corsairs, and Helldivers, forms up along Runway Five on May 2, 1945. (USN/Webster)

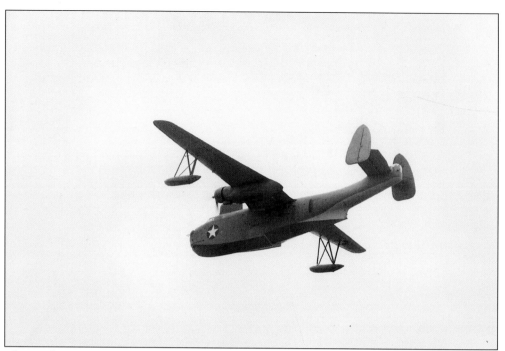

This radar-equipped ASDEVLANT PBM-3D Martin Mariner was photographed on July 8, 1943. (USN/Haus)

A New London-based fleet boat lays her anti-aircraft and main battery upon an ASDEV airplane off Block Island during exercises in 1943. (USN/Haus)

Starting in 1944, Quonset had a full-time PBY-5A-equipped Coast Guard Air-Sea Rescue Unit. (USCG)

Semper paratus and damn cold! The United States Coast Guard's one and only patrol squadron, VP-6(CG), was outfitted, trained for ASW, and supported by Quonset's Fairwing Nine. This photograph of VP-6(CG) PBY-5A (buno 08049) and her crew was taken at Bluie White One, Greenland, on January 15, 1944. Although ready and able to kill subs, Coast Guard Patrol Squadron Six's most valuable contribution was in ice and weather recon, plus search and rescue: a two hundred-year heritage. (USN/USCG)

The land plane side was busy as all heck in the summer of 1943. (NA)

The seaplane side was also photographed in 1943. Sub-hunting land-based Kingfishers and Lockheed PV Venturas dominate the ramp amongst a scattering of PBY and PBM flying boats, utility aircraft, and a lone PB4Y-1 Liberator. The Quonset huts stacked in front of Seaplane Hangar Two are line shacks. The brick structure before Seaplane Hangar One is the billet of HEDRON-9, the headquarters of Fleet Air Wing Nine. Mooring camels and a seaplane-refueling pipeline can be seen at the bottom right. (NA)

ASDEVLANT PBM-3D sits on her beaching gear between test and evaluation missions on the seaplane side in 1943. (USN/Haus)

This is what Quonset's hard work was all about in the Atlantic ocean: death to U-boats. A TBF from the USS *Card* (CVE-11) scores a direct hit and a near miss against a Nazi sub in August 1943. (USN)

Quonset's Assembly and Repair Department (A&R) overhauled many World War II planes, including PBYs. (USN/NARF Pensacola)

Douglas Dauntless SBD dive bombers from Quonset's first Essex Class carrier air group (CVG-16, USS *Lexington*) return to the air station after training exercises in 1943. Like most Quonset-trained carrier air groups, they would see heavy combat in the Pacific. (NA)

This is a September 9, 1944 photograph of Quonset's CASU-22 (Carrier Aircraft Service Unit), which helped operate NAAF Hyannis. (USN/QAM)

Wayne (DeWayne) Morris, a Hollywood actor and real Navy combat flying ace, shot down seven Japanese aircraft with Fighting Squadron Fifteen aboard the USS *Essex* (CV-9). He returned to Quonset in 1945 were he taught advanced fighter tactics. Other famous Californians that trained at Quonset included Dana Andrews, Hank Fonda, and Richard M. Nixon. Another former U.S. President, George Herbert Walker Bush, trained at Quonset and went on to become the youngest Navy combat pilot in the Pacific. (USN)

The USS *Block Island* (CVE-21) was photographed on October 12, 1943. This Rhode Island namesake knocked off seven U-boats before she herself was sunk, with terrible losses, during the Battle of the Atlantic. A new Jeep Carrier (CVE-106), that later operated from Quonset, was commissioned on December 30, 1944, and was named in her honor. (USN)

Although officially based in Norfolk, most of the Jeep Carriers that helped win the Battle of the Atlantic, such as the USS *Core* (shown here on May 20, 1944), came to Quonset to discharge or pick up aircraft. Training Quonset's pilots was common duty for the Jeep Carriers when they visited the air station. The Navy ran the guts out of these small, uncomfortable, but proud ships—Uncle Sam sure got his money's worth out of these. The *Core* killed six U-boats. (USN)

The "Old Crow," the USS *Croatan* (CVE-25), barely made 18 knots on January 14, 1944.

After being trained at Quonset in the use of rockets and other air-to-ground weapons, VOF-1 was deployed aboard the USS *Tulagi* (CVE-72) for the Invasion of Southern France (Operation ANVIL DRAGOON). On August 16, 1944, officers in the squadron's ready room listen with rapt attention as squadron mate gives the lowdown on early, and highly successful, operations against the enemy. The especially intense gentleman with the coffee cup is Task Group Commanding Officer Rear Admiral Calvin T. Durgin. After Corsica was secured, he went on to do great things in the Pacific. (USN)

By the summer of 1943, most of the seaplane side was taken over by land-based Lockheed medium-range patrol planes and Kingfisher Scouts of the Inshore Patrol Squadron. (NA)

One hundred and eighty fighters, night fighters, dive bombers, torpedo bombers, transports, and utility amphibian planes, plus three British squadrons, standby in front of Quonset's four land plane hangars. Nearly a hundred more are up training, while an equal number are undergoing routine upkeep and repair inside the hangars. During 1943, an average day for Quonset would include 1,200 takeoffs and landings, some better than others. Sometimes three or four days would roll by without a crash or major mishap. (NA)

These torpedo bombers, with wings folded up in front of A&R's massive hangar, are the rare Vought-designed Consolidated TBY-2 Seawolves. Considered lemons by most who got near them, especially experienced TBF pilots, Quonset's VT-154 nevertheless trained hard with them in preparation for the invasion of Japan that, thanks to the two Air Force nuclear strikes, never had to take place.(USN)

Quonset's VT-154, the one and only squadron completely equipped with Seawolves, is shown here in July 1945 putting in some formation work (the "K" on the tail meant Quonset). Although marked for VT-154 and operated by this torpedo squadron, Quonset's CASU-22 was the official custodian and maintenance head for all TBY-2s aboard. All have APS-6 radar provided by Quonset. (USN/Lawson)

TBY-2s were built at a converted Mack Truck factory in Allentown, PA. Although full of potential, Seawolves, like most new airplanes, were also full of bugs and were quickly junked by the Navy right after the war. Some of the last ones served with Quonset's Utility Squadron, VJ-15. (USN)

Project Yehudi TBM-3D, sporting sixteen spotlights, was evaluated by ASDEVRON at Quonset. The concept of the lights was to make aircraft invisible to Nazi U-boat watch standers and gunners. The crazy thing worked, but was just too weird for the Navy. (Grumman/Lovisolo)

A new PV-1 from Quonset's VB-125 (the former VP-82) gets a checkout over Narragansett Bay on July 25, 1943. This fine squadron alone killed four U-boats and damaged several others. (USN/Irish)

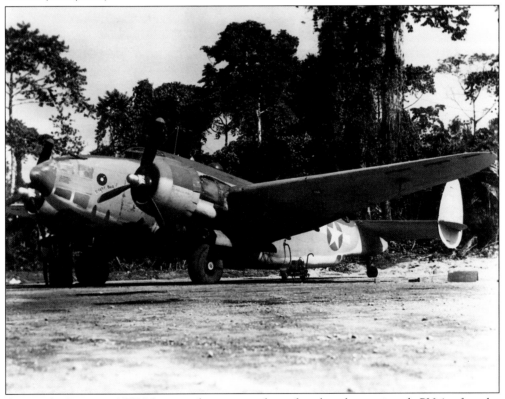

Quonset's Project AFIRM trained crews and outfitted radar-equipped PV-1s for the Navy's first night-fighting squadron, Marine VMF(N)-531. They got their first kill during November 1943, off Vella Lavella. This is a Gray Ghost PV-1 night fighter at Bougainville in late 1944. (USMC/Lawson)

Project AFIRM modified a few Twin Beach SNB-2Cs for evaluation. The APS-6 radar would eventually be used in night combat and ASW aboard Corsairs, Hellcats, and Avengers. (MIT Museum)

The Grumman F6F-3N, along with the F6F-3E, were the world's first naval night fighters built specifically for that purpose. This is a 1943 photograph. (Grumman/Lovisolo)

"Night Fighter Town USA!" The Navy's first night fighter was born in Rhode Island on April 3, 1942, in Quonset's Land Plane Hangar Three. Only eighteen months later, Navy night fighters began racking up kills in the Pacific. Quonset's VF(N)-75 was the first when one of its F4U-2 Corsairs splashed a Japanese Betty bomber off Vella Lavella, New Georgia, on October 31, 1943. All of the Navy's night-fighting squadrons got their start in Rhode Island, either at Quonset or at NAAS Charlestown. Two Quonset sailors are shown here checking the radar of a Grumman Hellcat. (NA)

In addition to supporting top-secret projects aboard the air station proper (such as AFIRM, ASDEV, SAIL, SINGER, and the TENTH FLEET), Quonset also supported many other projects at the Bedford (MA) Army Air Field, including BAKER, CADILLAC, CAST, and ROGER. In this 1945 photograph of the Bedford Hangar, Navy aircraft supported by Quonset included a Taylorcraft LNT-1 glider, a Martin JM-1 Marauder (buno 75194), at least seven SNB-2Cs, an FM-2 Wildcat, a Brewster SB2A-4 Buccaneer, and two Project CADILLAC TBM-3W2 Guppies. (MIT Museum)

"BEWARE OF ROTORS," stenciled in yellow on the blue/gray torque boom of a Navy HNS-1 helicopter, was a sign of things to come in naval aviation. Initially flown by Coast Guard pilots, HNS-1 helicopters first flew at Quonset in late 1944. The first one threw a chunk of a rotor blade into the head of a senior chief during a demonstration. The senior chief, with one hell of a headache—and one hell of an opinion of rotary-wing aviation—bullied his way out of sick bay the next day. In 1946, Quonset's NATU became the first Navy air unit to operate, full-time, this early helicopter. After wearing it out chasing torpedoes, it was donated to the Smithsonian in 1948. This photograph was taken on April 11, 1944, at South Weymouth. (NA)

In addition to training thousands of U.S. Navy pilots and aircrews during World War II, Quonset checked out hundreds of British personnel from more than a dozen squadrons in new, Lend-Lease aircraft, like this American-built Vought Corsair MK II. The "new boys," both British and American, found the Corsair to be a handful during initial carrier-landing qualifications, as this Royal Navy Number 1836 Squadron pilot found out in 1943! (Royal Navy/Curry)

The USS *Bogue* (CVE-9), with twelve U-boats sunk, was the top-scoring Jeep Carrier of World War II. Hunting and killing U-boats in the Atlantic was dangerous work, almost as dangerous as landing on this carrier was for some young Quonset TBM aviators off Block Island on July 26, 1944. Unsung heroes of the Navy, enlisted deck monkeys, airdales, and V Division-types, demonstrated great skill, saving from the deep-six what is left of two Avengers. (USN)

Showing hundreds of tons of ice formed by sea spray, the USS *Core* takes a breather from the wintery North Atlantic just after berthing at Quonset's Carrier Pier on January 25, 1945. By this time, she had killed five U-boats, and was due to chalk up another sub in April, one of the last ones destroyed before V-E Day. (NA)

Later to become "The Jet Admiral" and Quonset's tenth commander, Captain Daniel V. Gallery poses aboard the shot-up conning tower of the German submarine U-505 that his USS *Guadalcanal* (CVE-60) Task Group 22.3 tracked down, sighted, attacked, boarded, and captured on June 4, 1944. (USN)

This crew (consisting of a radioman, pilot, and turret gunner) belonged to the sub-killing "Len Sharon," a Grumman Avenger 58-C-26 from Quonset-trained VC-58. Composite Squadron Fifty-Eight killed three U-boats while working from the USS *Guadalcanal* (CVE-60) and the USS *Wake Island* (CVE-65). (Grumman)

In order to provide target aircraft for the top-secret Section T proximity-fuse anti-aircraft shell program, A&R Quonset converted just about every SBU-1 and SBU-2 Vought scout bomber that the Navy had into radio-controlled drones, using Singer Sewing Machine components. After modification, these handsome ships were immediately flown to Cape May, where they were quickly expended, proving the validity of the VT proximity fuse, which saved many ships and men off Okinawa during Kamikaze attacks. This photograph, taken at an earlier time, shows SBU-1s from the *Ranger*'s Air Group getting set for a 1937-style, three-plane-section takeoff. (Larkins)

Although traditionally composed almost exclusively of Marines, many sailors acted as security personnel after Marines were called back to the Pacific after Tarawa. (USN/Macia)

Two escort carriers (CVEs) load F6F Grumman Hellcats for carquals (Carrier Qualification deck landings), as nine destroyer escorts (DEs) nest alongside the Carrier Pier on June 22, 1945. When not killing Nazi U-boats, the highly-efficient Jeep Carriers trained aviators at NAS Quonset Point. (NA)

Still wearing a cast and suffering from more than one hundred shrapnel wounds taken when a Japanese suicide plane hit the USS *Ticonderoga* (CV-14), Quonset's beloved captain, Commodore Dixie Kiefer, was all smiles when he announced on August 14, 1945, that Japan had surrendered. (NA)

Three

Peacetime

The Saipan Class light carrier USS *Wright* (CVL-49) shares the Carrier Pier with the long-hulled Essex Class USS *Leyte* (CV-32) early in 1950. The ole "Forty Niner" held her own with the best of the big ones, and gave the Navy good service, both as a carrier, and later as a command communications ship. Her sister ship, the USS *Saipan* (CVL-48), also operated from Quonset and qualified the Navy's first operational jet fighter squadron, Quonset's own Fighting Seventeen Able (VF-17A), in FH-1 Phantoms. (USN/ Lawson)

This wonderful line of Quonset aircraft photographed in 1945 during the first V-J Day Open House includes: a PB4Y-1 Liberator; a R50 Loadstar; a SNB-2C "Slow Navy Bitch"; a Cessna JRC-1 T-50 Bobcat (like Sky King's "Song Bird"); a Yellow Peril and a Duck that both sport pre-war yellow top wings; a Grumman Goose and Widgeon; a PBJ Mitchell; a JM Marauder; a XJD-1 Invader; and a BTD-1 Douglas Destroyer. (USN)

The favorite ride of Assistant Secretary of the Navy for Air John Nicholas Brown, a native of Rhode Island, was the Consolidated PB2Y-3R Coronado. This giant patrol flying boat was modified to transport VIPs by including Pullman Car-type accommodations and changing the inboard propellers from three blade-types to less efficient, but *quieter*, four blade-types. (USN/PC)

The Navy's largest flying boat, the record-breaking Martin JRM-2 "Caroline Mars," visited Quonset during late June 1949 with a load of Middies up from Annapolis, as part of the Naval Academy's first summer "air cruise." The air station's seadrome was always kept in top condition and could handle *any* boat. (NA)

The Navy's largest postwar blimp, the M-4 *Big Mike*, operated briefly with Quonset's NATU for torpedo experiments, and was temporarily moored at NAAS Charlestown. Severe icing on the big ship caused it to return early to NAS Lake. (NA/Webster)

The Navy's first real carrier-based jet fighter was the McDonnell FH-1 twin jet Phantom, R-103 of Quonset's VF-17A, attached to Carrier Air Group Seventeen. Although tiny, short-legged (note the fixed ferry tank under its belly—a common feature), and under-powered, the FH-1s were durable, handled fine, and brought the Jet Age to the Navy's carriers, just in the nick of time. (Kirby)

The *Saipan* is shown here loading VF-17A FH-1 from the Carrier Pier for carquals in May 1948. Quonset was the home port for the world's first carrier-based jet squadron. (USN/ Lawson)

The Phantom's of Quonset's VF-17A did have some bugs, as relief tubes often backed up and the center-line bathtub fuel tanks often failed to separate when empty. (USN/Harris)

The Jet Age often arrived hard and deadly at Quonset. The residue of a cockpit section of a F9F fighter jet, deposited on the Carrier Pier by the *Maryann*, awaits pickup for transportation to the bone yard. (NA/Webster)

Big, powerful, capable, but no good around an aircraft carrier, the Martin AM-1 Mauler saw its first Navy duty with Quonset's VA-17A. Attack Seventeen Able wrote the book on the AM and found it severely wanting in elevator control. Armed with four 20 MM cannon, three 2,000-pound torpedoes, and twelve HVAR rockets, AMs initially "out-gunned" their rival, the early versions of the Douglas AD Skyraider, but not for long. This is a "pre-flight" of VA-17A AM-1 R-401 at Quonset in 1948. (USN/Lawson)

This cross-deck pendant fiddle bridge view shows Mauler R-407 (buno 22300) coming aboard the USS *Kearsarge* (CV-33) on December 27, 1948. Less-than-marginal elevator effectiveness doomed more than a few Mauler pilots at sea. (NA)

The AM could kill its pilots while coming in for shore landings, too. These are the remains of AM-1 (buno 22265), after it hit a bulkhead at the end of a Quonset runway on February 24, 1950. The Navy quickly moved the AM from carriers to the reserves and dedicated itself to Mr. Ed Heinemann's magnificent AD Skyraider series. The last fleet Maulers were two AM-1Q ECM birds that Comfair's VC-4 retired early during October 1950. (NA/Webster)

Viewed from the cab of a Quonset Marine's gray Chevy pickup, a rare and primitive Chance Vought F6U-1 Pirate jet fighter is secured on the seaplane ramp in 1949. Had these little ships gone into production, Quonset would have been their East Coast overhaul and repair headquarters. This one was destined to be an ordnance target. The three propeller-driven aircraft sharing the ramp area are early prototype AF-2S and AF-2W Grumman Guardians up from Bethpage for tests. (Leatherneck)

After extensive modification by Quonset's O&R for duty near the South Pole, ski and JATO-equipped R4D-5s blast off on January 30, 1947, from the flight deck of the USS *Philippine Sea*: destination Little America. In addition to skis and JATO, modifications to these ships included the installation of torpedo plane-type APS-4 radar under the starboard wings, an improved navigation station, improved electrical systems, insulation, and heating, as well as bright yellow rescue bands around the fuselage and wings. (NA)

Receiving finishing touches in front of O&R are four of the forty "DC-3s" that the big plant converted into R4D-7 "Flying Class Room" navigation trainers. All but one of the Navy's forty-one "School Ships" were products of Quonset, as were dozens of other R4Ds that were converted to R and Z configurations with plush airline-type and VIP accommodations. Still more R4Ds were converted to radio, radar, and ECM (Electronic Counter Measures) trainers and R&D platforms. Between 1946 and 1949, more than two hundred of the venerable "Goonie Birds" got O&R's full treatment. (USN)

Quonset's Stoner-Mudge fuel-tank system found its way into NACA (The National Advisory Committee for Aeronautics, the precursor of NASA), when it helped save the Navy's first Douglas D-558-1 Skystreak in 1947. (Dou6kins)

In his brown shoes and greens, reading Scripture in Quonset's Dixie Kiefer Chapel, Ensign Jesse L. Brown became the first black Navy aviator when he was designated in October 1948. The chapel was named for NAS's beloved captain, Commodore Kiefer, who perished with four others in a JRB Twin Beech on November 10, 1945, while returning from the Army-Navy game. (NA)

On a cold day over Quonset in November 1949, with the canopy rolled back and his flight goggles way up, Ensign Jesse Brown holds close formation on a camera plane with his VF-32 F8F-2 Grumman Bearcat, insuring the Harry S. Truman administration that integration was truly underway in the Navy. Brown was shot down in Korea on December 4, 1950; his VF-32 squadronmate, LTJG Thomas Jerome Hudner Jr., attempted to rescue him, but to no avail. America's first black naval aviator, and the first one to fire his guns in combat, was gone. Flying from USS Leyte (CV-32), Quonset's CAG-3 later dropped napalm on the crash site, consuming Brown's remains. Hudner received the Congressional Medal of Honor for his actions. (NA)

After a hard landing, the pilot of Quonset's VF-33 F8F-2 Bearcat hauls out of the burning cockpit. (He made it!) Bearcats were the best operational piston-engined fighter planes in the world, but they were not jets. The Navy's need for prop-driven attack and fighter-bomber types, such as ADs and Corsairs, which had greater range, weapons load, and loiter capability, sent the Bears to the reserves (and the French). (NA)

Adorned in apple-green zinc-chromate primer, F4Us and ADs pass through their respective production lines at O&R in March 1950. Many of these ships would see action in Korea. (NA)

On September 5, 1946, Navy Development Squadron Four (VX-4) changed its home port from NAS New York to Quonset. The PB-1W flying radar stations, converted from World War II Army Boeing B-17G Flying Fortress heavy bombers, were equipped with large APS-20 radar domes under their bomb bays. The main task for VX-4 at Quonset was ASW and Airborne Early Warning Development (AEW), both done perfectly under their first CO, the highly-competent Commander Lucien F. "Red" Dodson. (Bowers collection)

This photograph shows a continuation of the World War II Project CADILLAC: part of VX-4 PB-1W CIC (Combat Information Center), in a former B-17's radio compartment. (MIT)

Long after its eighteen months of duty at Quonset with VX-4, Boeing PB-1W (buno 77234) starred in the 1961 World War II movie classic about the Eighth Air Force, *The War Lover*. It appeared, complete with broom-handle .50 caliber guns in the top turret, as Steve McQueen's ship "The Body," B-17G DF—V. (Peter M. Bowers collection)

Quonset's flying sailors off Block Island helped develop the Navy's early radar picket submarine efforts, including Project MIGRAINE's first real one, the USS *Spinax* (SSR-489). The terrible loss of surface small boys as pickets against Japanese Kamikaze suicide planes spawned this postwar effort that continued into the Cold War. (USN/USNI)

Quonset's aviation assets helped with the evaluation of captured ships, such as the highly-streamlined Type XXI Nazi EX-U-3008 advanced-design snorkler, shown here on August 30, 1946. The USSR also carefully studied their German hi-tech spoils of war, and quickly developed an undersea threat to the world's democracies that Quonset would defend against for the rest of her operational life as a U.S. Naval Air Station. (NA/NHC)

Quonset's military and civilian personnel apply mechanical fog and chemical foam to the flaming tail section of a crashed AD Skyraider. (NA/ Webster)

Ready for "anything," with all manner of prepared slings, bridals, pelican hooks, and saddles, Quonset's homemade "Tilly" prepares to lift remains of a crashed TMB-3E. (NA/Webster)

Quonset's "Hound Dogs" coming home: these are P2V-2 Neptunes of VP-8 just clearing Newport while returning from exercises in 1950. (USN/VP-8 Alumni)

Officers of Quonset's Fleet Air Wing Three pose on the eve of the Korean War before one of their charges, a freshly-minted Lockheed P2V-4 Neptune, sporting six 20 MM nose guns and APS-20 radar. (NA)

U.S.S. Cabot
CVL 28
Entering Havana Cuba
February 8, 1949

A real Pacific War veteran, the USS *Cabot* (CVL-28) operated from Quonset as an ASW carrier, a training carrier, and performed many research and development tasks. Seen with CAG-11 aboard, the "Iron Lady" enters Havana, Cuba, on February 8, 1949. (USN/Bill Mulholland)

Corsairs from CAG-4 and CAG-10 await induction into Quonset's O&R Plant on October 20, 1949. They stand behind ADs and F4Us in primer coats being reassembled. (USN)

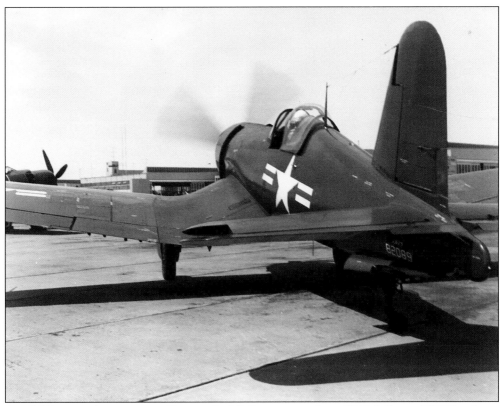

An O&R test pilot returns a "factory fresh" Quonset product, F4U-4 (buno 82089), after a check flight. The skill and dedication of Quonset's O&R family was known worldwide. (USN)

Out of a cold Narragansett Bay mist, a Comfair Quonset HU-2 Piasecki HRP-1 Flying Banana UR-9 tandem helicopter flares for a landing. The flight deck of the USS *Leyte* has a horde of VA-17A AM-1 Martin Maulers on it in this January 9, 1949 photograph. (NA)

Sikorsky H03S-1, one of two assigned to Quonset in the late 1940s, demonstrates rescue-hoist techniques to civilians in 1949. Worth their weight in gold during the Korean War for combat aircrew rescue, Quonset turned over both of her Sikorskys (Q-8 and Q-9) to the Pac. (USN/QAM)

In this April 1949 photograph, Quonset's VA-35 loads its TBM-3Es for deployment at the Carrier Pier. During the Truman administration, Douglas AD Skyraiders were in short supply, so venerable but obsolete World War II torpedo planes had to fill in. (NA)

Mounted in his F6F-5 Hellcat in 1947, the skipper of Quonset's Carrier Air Group Three displays his "Jet Age" hardware, mostly of World War II vintage: a Helldiver, an Avenger, and a Bearcat. Under President Harry S. Truman, most aviation appropriations went to the newly-created, A-bomb-slinging United States Air Force. (USN/Lawson)

Disaster! An explosion and the resulting five million-dollar fire enveloped Quonset's O&R Engine Shop shortly after 3:00 pm on October 15, 1948. All hands, Navy and civilian, instantly turned to rescuing dozens of trapped personnel. (USN)

Looking like a scene from Pearl Harbor, partially-completed Project ROGER R4D is saved from the onrushing flames of the October 15, 1948 fire. (NA)

Nearing completion on the site of the big fire, the new O&R Engine Division Building, designed from the ground up for piston and jet-engine overhaul, was the first of its kind in the Navy. This photograph, taken on April 21, 1950, shows hordes of ADs and Corsairs awaiting induction, as well as nine R50 Loadstars abaft Runway Five. (NA)

After World War II, Quonset's A&R (Assembly and Repair) was renamed O&R (Overhaul and Repair), as the days of "assembling" biplanes from piano crates supplied by aircraft manufacturers was long over. The F4U Corsair and Douglas AD Skyraider line, in full production, broke many Navy records and received many awards. The Quonset civilian work force always contained many veterans, so finished products were top-notch. (NA/Webster)

The busy Quonset Aircraft Modification Unit in Seaplane Hangar One upgrades a CAG-3 VF-31 F9F-2 Panther, an EX VF-171 FH-1 Phantom, plus dozens of Navy and Marine Corsairs and ADs in March 1950. The small rectangles near the tail pipe of the Panther are photographs depicting changes to be carried out. (NA)

A lone SNB-2C shares Quonset Aircraft Modification Unit spaces in Seaplane Hangar One with active and reserve Corsairs, Marine Corsairs from Cherry Point's HAMRON-13, plus ADs, Bearcats, and F4Us from Quonset's own CAG-3 and CAG-7. "MARINE AIR RESERVE NORFOLK" is stenciled on the nose of the Corsair with the reserve stripe in the foreground of this March 1950 photograph. (NA)

Gleaming, "factory fresh" Corsairs and ADs receive the latest upgrades by Quonset's Aircraft Modification Unit in March 1950, before being issued to the fleet. Quonset always meant *quality*. (NA)

The Navy's Construction Battalion, the Seabees, got their start at Davisville, literally, just up the street from Quonset in 1942. This bee, complete with snail-drum Thompson submachinegun, wrenches, mallet, pry bar, claw hammer, and attitude, was a Quonset landmark. Take a right at the bee and in one minute you are at Quonset Point's Main Gate. (USN/PC)

With sandbags removed and stripped-awnings rigged for NATS (Naval Air Transportation Service) passengers, Quonset's Control Tower/Operations Building puts on a nice fresh face on March 14, 1946. (USN/Irish)

The USS *Leyte* and the USS *Randolph* (CV-15), with CAG-7 and CAG-17 aboard, share the Carrier Pier on January 31, 1947. (NA)

Quonset's Utility Squadron Five (old VU-5) provided all manner of targets to shoot at. This Douglas JD-1 Jig Dog U-E-1 (the Navy version of the Army A-26 Invader light bomber) could drop target drones as well as drag sleeves and banners. This photograph was taken on April 16, 1947. (NA)

Quonset's CAG-7 Grumman F8F-2
Bearcats aboard the "Phil Sea"
(CV-47) were photographed off
Greece in March 1949. This excellent,
newly-minted, long-hulled, Essex
Class aircraft carrier languished at
Quonset for more than twenty months
waiting for a full-time crew while
under the Truman administration.
On the eve of the invasion of South
Korea by North Korea, the entire
United States Navy possessed only
six operational full-size aircraft
carriers. Three of them, with their air
groups, were homeported at Quonset.
Two would fight in Korea. (NA)

During the 1949 "Admirals' Revolt,"
Quonset almost got the chop from the
newly-formed Defense Department,
which favored large "strategic bombers"
over the proven aircraft carriers.
Lines of ten-engined Convair RB-36F
"Peacemakers," at 3.2 million dollars
a crack, rolled out, day and night, at
Fort Worth. (General Dynamics)

Then and now, this is the world's most majestic weapon: an American aircraft carrier. Seen from Pri Fly's roost aboard her sister ship, the USS *Kearsarge* (CV-33), the Leading Lady, the USS *Leyte* (CV-32), is swung by commercial tugs in the turning basin prior to berthing at Quonset's Carrier Pier on March 22, 1949. (NA)

This photograph shows the last peacetime Christmas that NAS Quonset Point would ever see. The USS *Philippine Sea* (CV-47), all tricked out for the holidays, was at the Carrier Pier in December 1949. (NA)

Four

Korea

With Task Force 77, rocket-armed Corsairs from Quonset's CAG-3, plus a few F4U-5N night fighters from Comfair's NAS Atlantic City-based VC-4, stand by for another wartime launch from the USS *Leyte*, as savvy "deck apes" remove snow and ice with steam hoses on December 13, 1950. (NA)

Plane handlers on the *Leyte* push a Quonset CAG-3 F4U-4B Corsair into the hangar deck from the beam elevator after a strike on enemy positions in Korea on November 11, 1950. The *Leyte*'s Air Group was among the first three to see early action in Korea. (NA)

A fleet oiler pumps Bunker-C, AVGAS, and JP Jet Fuel to the thirsty *Leyte* while also tending to the needs of a Gearing Class destroyer, the USS *Henderson* (DD-785), off Korea on November 16, 1950. Fleet oilers and other Navy auxiliary ships, then and now, make carrier task forces and carrier battle groups viable, worldwide, every day of the year, forever. (NA)

Bombed-up ADs and rocket-armed Corsairs from Quonset's CAG-3 are readied for launch from the *Leyte* against Korean targets on November 21, 1950. (NA)

Back from a Korean air strike, with all ordnance expended, one of Quonset's VA-35 "Black Panther" AD-3s taxis to the forward stack aboard the *Leyte*. CAG-3 in-chopped Korea on October 9, 1950. (NA)

These CAG-3 VF-31 Grumman F9F-2 Panther jet fighters were photographed aboard the *Leyte* in 1951. (NA)

Aboard the "Bonnie Dick" (the USS *Bon Homme Richard*, CV-31), Quonset's CAG-7 VF-72 Panther jets get their 20 MM guns checked, while red-shirts load bombs in preparation for the squadron's first Korean War strike on June 23, 1951. (USN)

Quonset's VP-7 P2V Neptunes went to Korea and served in maritime recon and ASW roles with Task Force 77. Fortunately, they were not bounced and chewed up by 23 MM and 37 MM cannon-armed, Russian-built and piloted MiG-15 jet interceptors. This August 24, 1950 photograph, taken at Quonset, shows a Neptune 20 MM tail gun turret. (NA)

The chief petty officer of Quonset's FAW-3 Patrol Squadron Seven (VP-7) insures that 260-pound live depth charges find their proper spot in the bomb bay of a P2V-4 Neptune on August 24, 1950. (NA)

One of Quonset's VA-75 "Sunday Punchers" AD-4s gets picked up from a barge in Wonson Harbor by the battleship USS *Iowa* (BB-61) in September 1952. (NA)

Launch jets! Grumman F9F Panthers are catapulted from the *Leyte*. Almost half of the Leading Lady's Korean combat missions were jet sorties. (USN)

Within hours of confirmation that North Korea had invaded South Korea, Quonset's Marines buttoned up the air station and screwed her down tight with all manner of savvy security measures that included random spot checks of vehicles and personnel. In 1950 this civilian-registered Mercury sedan was stopped by a Quonset Marine corporal mounted in a half-ton Chevy pickup truck (USN 287213) sporting a giant, manifold-driven siren on its left front fender and a Quonset Scout decal on its lower left door. (Leatherneck)

One of the best postings in the Corps was the Marine Barracks, U.S. Naval Air Station Quonset Point, RI. The First Dog Watch at the main gate was photographed in 1951. (Leatherneck)

Quonset's CAG-3 Panthers and Corsairs are shown here aboard the *Leyte*. (USN/Curry)

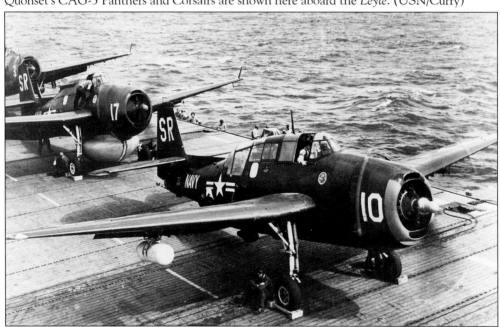

While Quonset's attack carrier air groups were fighting in Korea, its ASW squadrons fought the Cold War Battle of the Atlantic against Russian submarines. VS-32 prepares to launch a TBM-3S Striker and a TBM-3W2 Guppy from the USS *Salerno Bay* (CVE-110) on March 2, 1952. Quonset's VS-32 was one of the lucky few ASW Squadrons that never flew AF-2 Guardians. VS-32 stayed with Teebs (the Grumman-designed, General Motors-built TBMs) until transitioning to twin-engined S2F-1 Trackers. (USN)

A VS-32 APS-20 radar-equipped TBM-3W2 Guppy launches from the *Salerno Bay* during Atlantic Fleet Exercise CONVEX III, on March 2, 1952. (USN)

With a bone in her teeth, Quonset's *Salerno Bay* respots searchlight-equipped TBM-3S Strikers and TBM-3W2 Guppies as she replenishes a Gearing Class destroyer while on exercises in the Caribbean on March 5, 1952. (NA)

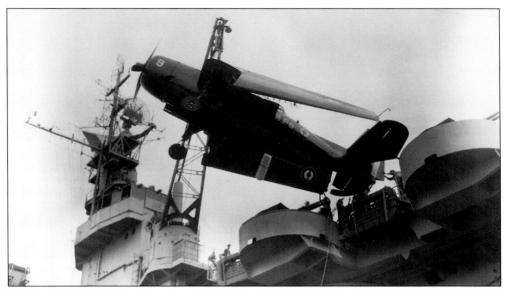

Bon voyage en bon chance! From the late 1940s through the mid-1950s, much of America's support of French naval aviation, under the Military Assistance Program (MAP), was provided by Quonset. This included training aircrews and technical personnel, plus the overhaul, upgrading, and detailed modification (for service in the French Navy) of USN fighter and attack types, all of Grumman-design. In this July 17, 1951 photograph, an airplane hoist of the French light carrier *Lafayette*, formerly the USS *Langley* (CVL-27), lifts aboard a TBM-3E Avenger at Quonset's Carrier Pier. (NA)

Guppy and Striker VS-31 Guardians pose as a Hunter/Killer ASW team over Quonset. VS-31 maintained, along with other Guardian squadrons, a ready alert team that, on thirty minutes notice, would head out in pursuit of Russian submarines. They only carried 20 MM gun loads and flares; HVAR rockets, ASW depth charges, sonobuoys, and torpedoes stayed ashore. These were the rules of the road in the Cold War Navy. (USN/Lawson)

Five

The New Navy

Unlike previous Avenger and Guardian-converted torpedo bombers, this "hi-tech" Grumman S2F-1 Tracker was built for the ASW job from the ground up. In this 1958 photograph, one of Task Group BRAVO's VS-31 Top Cats awaits cat shot from the USS *Tarawa* (CVS-40). Unlike earlier carrier-based aircraft, "Stoofs" performed both the "hunter" and the "killer" missions, which had previously required two aircraft. (USN/Quonset Scout)

Wrecked, damaged, and swamped Stoofs, Panthers, and Corsairs were photographed on the seaplane side after taking a direct hit from Hurricane Carol's high winds and storm surge on August 31, 1954. (NA)

Built with an enlarged bomb bay and a bigger tail in order to deliver Weapon Betty (a fat, low-tech nuclear depth charge), a Quonset-based VS-31 S2F-2, with ditching hatches open and still in blue paint, is readied for launch. More compact nukes rendered this well-thought-out ship quickly obsolete, but the big tail stuck with all future S2F production designs. Many of the sixty production S2F-2s were usefully converted to US-2C target tugs at NARF Pensacola early in the Robert Strange MacNamara era. In new gray paint, a much more proliferate S2F-1 follows. (USN/Curry)

Quonset's plane handlers and catapult crews were the unsung heroes at sea, performing some of the most dangerous jobs in the world. These crews of the USS *Tarawa* (CVS-40) prepare VS-31 S2F-ls for launch from "Building 40" amid deadly, close-coupled, whirling twin props that were a new phenomenon aboard ASW support carriers in 1955. (USN)

America's (and the world's) second nuclear-powered submarine, the USS *Seawolf* (SSN-575), is greeted in Long Island Sound by a Quonset-based HS-9 HSS-1 ASW helicopter. Quonset's fixed-wing and rotary-wing squadrons frequently worked ASW problems and exercises with early New London-based nukie boats that also included the very first one, the USS *Nautilus* (SSN-571). With a single-piston engine and no auto-hover system, "Hiss ONE" helicopters were a handful, especially at night while dipping sonar—tough, dangerous, usually thankless work.

Quonset's VA-72 was the first squadron to receive the ingenious Douglas A4D-1 Skyhawk bantam tactical nuclear bombers designed by "Mister Attack Aviation" himself, Ed Heinemann. This photograph was taken on October 30, 1956. (NA)

The reason for the Skyhawk's long landing gear is evident when nuclear practice shape is loaded on an early A4D-1. Most Rhode Islanders had no idea of Quonset's nuclear capabilities. (Douglas/Gann)

Carrier duty, then and now, in war and peace, is always dangerous. After an explosion and fire at sea, the USS *Bennington* (CVA-20) transfers more than one hundred injured personnel to rows of gray ambulances, a med-evac bus, and an amphibious landing vessel up from Newport at the Carrier Pier. Coast Guard H04S-3G and Navy HU-2 HUP-2 helicopters also lifted injured shipmates directly from Big Ben's flight deck to the Newport Naval Hospital on May 26, 1954. (NA)

Rear Admiral John R. Hoskins was Comfair Quonset from March 12, 1954, to September 19, 1956. During World War II, he served on the light carrier USS *Princeton* (CVL-23), where he lost a leg due to a Kamikaze attack on October 24, 1944. Despite this, Hoskins managed to qualify in jet fighters, and got command of the new big USS *Princeton* (CVA-37). The movie *The Eternal Sea* is based on his inspiring life story, and had its world premier in Providence during the fall of 1955. (USNI PC)

Half of the Navy's entire fleet of TF-1Q Grumman Electronic Counter Measures aircraft belonged to Quonset's VA(AW)-33 Night Hawks. Heavily-modified CODs had Stoof wings and plied their still-classified trade up and down the East Coast for more than a decade. (USN/Curry)

The Navy's first functioning carrier-based nuclear attack squadron VF(AW)-4 (formerly VC-4) came home to Comfair Quonset in 1959 and was soon disestablished. With an all-weather and night-fighting legacy going all the way back to Quonset's World War II Project AFIRM, the squadron turned over four of its F2H-4 jets to Quonset's Utility Squadron, VU-2, where they served briefly as the last fleet Banshees.. (USN/Lawson)

VU-2 shows off its new TV-2 on September 8, 1959, while FJ-3 Fury jets and a JD-1 Jig Dog tow plane provide a background for the KD2R-5 target drone. VU-2's KD-Unit rode Newport-based destroyers and cruisers, catapult launching the small, expendable drones for target practice. (USN/Brooks)

A VC-4 AD-5 throws a flaming, 2,000-pound engine down the flight deck after a late wire hard landing. Atlantic and Mediterranean Fleet Navy and Marine AD Squadrons supplied Quonset's O&R with plenty of work for more than two decades. (USN/Lawson)

Rated as the top U.S. Industrial Naval Air Station, Quonset's O&R employed more than 5,000 civilian workers, mostly Rhode Islanders, when this photograph was taken on February 12, 1958. (USN)

A HU-2 HUP-2 from the USS *Block Island* (CVE-106) rescues a VS-22 pilot from his sinking AF-2 after a failed takeoff on August 12, 1953. (USN/Lawson)

History was made in September 1957, when Marine Helicopter Squadron One (HMX-1) performed the first Presidential lift with the original Marine One helicopter, a Sikorsky HUS-1Z. Dwight David Eisenhower flew from the Summer White House in Newport to Quonset when he was required to return to Washington on short notice. (USMC/HMX-1)

The USS *Wasp* (CVS-18), with an unusual detachment of six Grumman F9F-8 Cougars, is shown here near Quonset's Carrier Pier. The six smart-looking vessels nesting at the Carrier Pier are destroyers of the Royal Canadian Navy (RCN). During the Cold War, the Quonset command was the locus of all ASW Hunter-Killer (HUK) efforts with Her Majesty's Canadian ships and aircraft. Joint exercises like this were common and fruitful in the 1950s and sometimes included the Canadian light carrier HMCS *Bonaventure*. (USN)

More than four thousand first-shift Quonset civilian employees hit the road for the daily man-to-man, bumper-to-bumper, battle on Route One in 1951. Despite efforts to convey employees to and from Quonset by bus, rail, and boat, single-occupant Fords, Chevys, Dodges, and Plymouths were always the norm. (USN)

No fewer than thirty-eight Douglas AD Skyraiders, and one lone little F6F-5 Grumman Hellcat, are de-fueled as they await paint-stripping and induction into Quonset's massive O&R plant on February 12, 1958. Quonset was the repair head for all ADs on the East Coast. The hangar at the bottom left is the location of the Quonset Air Museum. (USN/Quonset Scout)

114

Six

The Sixties

The "Champ," the USS *Lake Champlain* (CVS-39), was the last of her kind and served at Quonset through the late 1960s, hunting Russian subs and picking up NASA astronauts. (USN/Curry)

In addition to preparing attack, AEW, and ECM Skyraiders for duty worldwide, Quonset provided hundreds of AD-4s, AD-5s, and AD-6s to the U.S. Air Force and the Republic of Vietnam Air Force. By August 1967, Quonset's O&R strictly dealt with ADs, Stoofs, and WF-2 "Fudd" radar planes. All modern, jet-type production went south, to Virginia and Florida, as would all of Quonset's other assets, six years later. (USN/Thomas F. Hannan)

Many Skyraiders that were overhauled at Quonset served the Navy well during the Vietnam War. This AD-6 from the USS *Coral Sea* (CVA-43) launches with a full load of ordnance for a combat mission, c. 1967. (USN/Rausa)

Quonset's "Miss T" is shown here at sea off Gitmo on July 14, 1957. (USN/USNI)

The "Champ" was photographed in her ASW prime, with plane guard HU-2 HUP, VS-22, and VS-32 Stoofs, VAW-12 AD-5W Guppies, and HS-5 Hiss One Nans. (USN/Curry)

This Night Hawk was photographed above the weather. After VAW-12 departed for Norfolk, VAW-33 began to supply AD-5W DETs to all East Coast ASW carriers. (USN/Curry)

VAW-33 flew some of the last Navy combat Skyraider missions in Vietnam with ECM configured EA-1Fs (AD-5Qs). During the early 1960s, VAW-33 flew all four Navy multiplex-type Skyraiders, AD-5N Nans, Ad-5Q Queens, AD-5W Guppies, and a COD-configured AD-5. This is the USS *America* (CVA-66) detachment. (USNI/Curry)

The Navy's first operational Grumman WF-2 Tracers came to work at Quonset's VAW-12 on January 20, 1960. Two years later, the squadron moved its Fudds to Norfolk. (USN)

This Grumman S-2E Tracker, CVSG-52, is from the USS *Wasp* (CVS-18). (USN)

Hold that hover! A Quonset HS-9 Sea Griffin HSS-1 ASW Seabat helicopter hovers over a nasty, as usual, North Atlantic. (USN/Curry)

A sub-hunting HS-11 SH-3 makes a low pass over Marines at Quonset's Main Gate. (USN/Browning)

Always the big draw at Quonset's open house activities and Navy relief carnivals, the Blue Angels Flight Demonstration Team (later Squadron) is shown here in their second phase, Grumman F11F-1 Tigers. (Lawson)

A hippie in the 1960s examines Comfair Quonset's spit-and-polished, VIP-configured, dark gray Sikorsky helicopter during one of the air station's ever popular open house presentations. In addition to duties as the Admiral's Barge, the VH-34D lifted top American and foreign brass to the Naval War College in Newport as well as other primary destinations. (Serif/Author)

Dedicated to the exploration of the Antarctic and the support of Operation DEEP FREEZE, Quonset's VX-6 operated a variety of fixed and rotary-wing machines. The Quonset-based VX-6 R4D-5L named "Que Sera Sera" made the first landing at the South Pole during operation DEEP FREEZE in 1956. She now resides in Pensacola NAS. When not summering upon South Polar glaciers, supporting helicopter operations, UC-1 De Haviland Otter JD-15 dropped squadron parachutists over the Great Rhode Island Swamp. (USN/VX-6)

Quonset's FAD maintained a fleet of SNB-5 "Bug Mashers" as proficiency machines so that students and staff at the Newport Naval War College could stay on flight skins. They were replaced by older Stoofs in the late 1960s. (USN)

This photograph shows how impressive NAS Quonset Point truly was. (USN/QAM)

Grumman A-6A Intruder all-weather jet bombers and Douglas A4D jet Skyhawks insured the survival of the great naval air station until August 1967, when local political negligence saw production of both planes head south to Virginia and Florida. From this point onward, all aircraft production at O&R (NARF) was piston-engined and propeller-driven: effectively a dead end, as would be seen six years later when the air station was ordered closed. After being completely overhauled at Quonset, jet A-6As like this one would receive a complete pre-flight check and a test flight by Quonset NARF test pilots before being sent to the fleet. (USN/ Quonset Scout)

The last aircraft assigned to the air station, Quonset's unique Grumman twin-tailed F-1 (C-1A) COD departs on April 3, 1974, for its new home port: NAS Norfolk, Virginia. Quonset's last skipper, Captain Edward J. Klapka, made the final official air station landing. The first air station landing was made by Quonset's first skipper, Commander Harold J. Brow, in another Grumman COD, a J2F-1 Duck, more than thirty-three years earlier. (USN)

Seven

Denouement

"Hello Quonset—Good Bye Quonset." The ship's company of the USS *Wasp* mans the rail, spelling out the name of the great naval air station, while reporting for duty, in earlier, happier times. Quonset, considered by many to be the finest industrial naval air station in the world, officially died on April 5, 1974. (USN/PC)

An A-4E Skyhawk from Quonset's USS *Intrepid* (CVS-11) intercepts a giant Russian Tupolev TU-20 Bear-D long-range strategic reconnaissance bomber in 1973. Big Bear-Bs, armed with AS-3 missiles, were a real threat to America and her aircraft carriers. The roar of the huge transonic counter-rotating propeller turbines could actually make Scooter pilots ill if they got too close while exchanging digital pleasantries with Russian turret gunners. (USN/Curry)

After visiting his World War II alma mater, President Richard M. Nixon ordered the air station closed. The Navy chop list was extensive, and as direct U.S. participation in Vietnam wound down, the Navy completed its shut down of offensive airborne anti-submarine warfare and went totally on the defensive against Russian nukie boats. Eventually, all of the Navy's ASW CVS support carriers and their air organizations above the squadron level would be scrapped.

Quonset employed 5,124 civilians, 4,822 of them Rhode Islanders, as of January 1, 1970. Three years later, all jobs would be lost. (USN/PC)

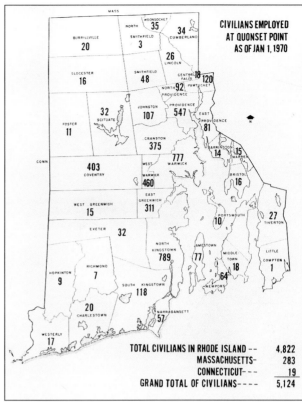

CIVILIANS EMPLOYED AT QUONSET POINT AS OF JAN 1, 1970

MASS.

WOONSOCKET 35
NORTH SMITHFIELD 34 CUMBERLAND
BURRILLVILLE 20
NORTH 3
LINCOLN 26
SLOCESTER 16
SMITHFIELD 48
CENTRAL FALLS 18
PAWTUCKET 120
NORTH PROVIDENCE 92
FOSTER 11
SCITUATE 32
JOHNSTON 107
PROVIDENCE 547
EAST PROVIDENCE 81
CRANSTON 375
WARWICK 777
BARRINGTON 14
WARREN 15
COVENTRY 403
WEST WARWICK 460
BRISTOL 16
WEST GREENWICH 15
EAST GREENWICH 311
PORTSMOUTH 10
TIVERTON 27
EXETER 32
NORTH KINGSTOWN 789
JAMESTOWN 77
MIDDLE TOWN 64
18
LITTLE COMPTON 1
HOPKINTON 9
RICHMOND 7
SOUTH KINGSTOWN 118
NEWPORT
CONN
CHARLESTOWN 20
NARRAGANSETT 57
WESTERLY 17

TOTAL CIVILIANS IN RHODE ISLAND -- 4,822
MASSACHUSETTS- 283
CONNECTICUT- - - 19
GRAND TOTAL OF CIVILIANS- - - - 5,124

Quonset's operational importance to the Navy during the Cold War is clearly illustrated in this 1972 operational control chart, when the commander of the Anti Submarine Warfare Force, Atlantic Fleet, hoisted his flag aboard the air station. (USN/PC)

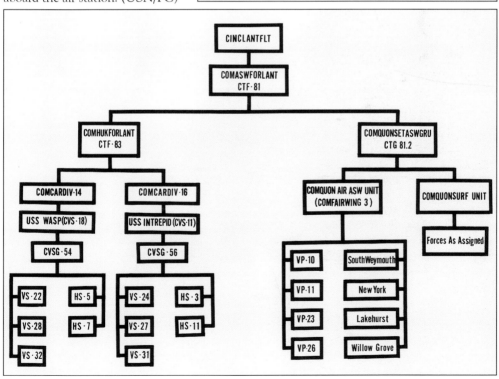

CINCLANTFLT

COMASWFORLANT CTF-81

COMHUKFORLANT CTF-83

COMQUONSETASWGRU CTG 81.2

COMCARDIV-14
USS WASP(CVS-18)
CVSG-54
VS-22 HS-5
VS-28 HS-7
VS-32

COMCARDIV-16
USS INTREPID (CVS-11)
CVSG-56
VS-24 HS-3
VS-27 HS-11
VS-31

COMQUON AIR ASW UNIT (COMFAIRWING 3)
VP-10 SouthWeymouth
VP-11 New York
VP-23 Lakehurst
VP-26 Willow Grove

COMQUONSURF UNIT
Forces As Assigned

This 1970s insignia indicates the air station's three major operational roles: Commander, Fleet Air; Commander, Air Bases, First Naval District; and Commander, Quonset Antisubmarine Warfare. (USN/PC)

Many of Quonset's over 5,000 civilian workers believed that Quonset, as had been announced earlier, would be the East Coast Naval Aircraft Rework Facility Head for the new, turbo-fan sub-hunter, the S-3A Lockheed Viking. Sadly, this was not the case. (USN/Quonset Scout)